Copyright 2024 by Deanne Dietz
All rights reserved. No part of this publication may be reproduced,
stored or transmitted in any form or by any means, electronic, mechanical,
photocopying, recording, scanning, or otherwise
without written permission from publisher.
It is illegal to copy this book, post it to a website,
or distribute it by any other means without permission.

Physical Drawings Converted to
Digital Drawings by: Moch. Fajar Shobaru from Fiverr - @mfShobaru

To Connect with Author Deanne Dietz:
Website: www.authordeannedietz.com
Instagram: deanne.dietz.author
TikTok: deanne.dietz.author

THE TEENY TINY TELESCOPE FOUND THE BIG BELT QUITE FUN,
AND JUMPED FROM LARGE TO SMALL ASTEROID ALL AROUND THE SUN!

THE TEENY TINY TELESCOPE FOUND SPACE TO BE QUIETER THAN THE DARKSIDE OF THE MOON.

THE TEENY TINY TELESCOPE DIDN'T KNOW WHAT TO DO.

THE TEENY TINY TELESCOPE SPOTTED COMETS APPROACHING BY AND THOUGHT THEY MAY BE ABLE TO LEAD THEM ACROSS THE SKIES!

www.ingramcontent.com/pod-product-compliance
Lightning Source LLC
Chambersburg PA
CBHW040351220526
45473CB00009B/2858